T0224952

# Cambridge Elements ☰

Elements in Geochemical Tracers in Earth System Science
edited by
Timothy Lyons
*University of California*
Alexandra Turchyn
*University of Cambridge*
Chris Reinhard
*Georgia Institute of Technology*

# THE CHROMIUM ISOTOPE SYSTEM AS A TRACER OF OCEAN AND ATMOSPHERE REDOX

Kohen W. Bauer
*University of Hong Kong*

Noah J. Planavsky
*Yale University, Connecticut*

Christopher T. Reinhard
*Georgia Institute of Technology*

Devon B. Cole
*Georgia Institute of Technology*

CAMBRIDGE
UNIVERSITY PRESS

# CAMBRIDGE
## UNIVERSITY PRESS

University Printing House, Cambridge CB2 8BS, United Kingdom

One Liberty Plaza, 20th Floor, New York, NY 10006, USA

477 Williamstown Road, Port Melbourne, VIC 3207, Australia

314–321, 3rd Floor, Plot 3, Splendor Forum, Jasola District Centre,
New Delhi – 110025, India

79 Anson Road, #06–04/06, Singapore 079906

Cambridge University Press is part of the University of Cambridge.

It furthers the University's mission by disseminating knowledge in the pursuit of
education, learning, and research at the highest international levels of excellence.

www.cambridge.org
Information on this title: www.cambridge.org/9781108792578
DOI: 10.1017/9781108870443

© Kohen W. Bauer, Noah J. Planavsky, Christopher T. Reinhard, and Devon B. Cole 2021

First published 2021

A catalogue record for this publication is available from the British Library.

ISBN 978-1-108-79257-8 Paperback
ISSN 2515-7027 (online)
ISSN 2515-6454 (print)

# The Chromium Isotope System as a Tracer of Ocean and Atmosphere Redox

Elements in Geochemical Tracers in Earth System Science

DOI: 10.1017/9781108870443
First published online: January 2021

Kohen W. Bauer
*University of Hong Kong*

Noah J. Planavsky
*Yale University, Connecticut*

Christopher T. Reinhard
*Georgia Institute of Technology*

Devon B. Cole
*Georgia Institute of Technology*

**Author for correspondence:** Noah J. Planavsky, noah.planavsky@yale.edu

**Abstract:** The stable chromium (Cr) isotope system has emerged over the past decade as a new tool to track changes in the amount of oxygen in Earth's ocean-atmosphere system. Much of the initial foundation for using Cr isotopes ($\delta^{53}Cr$) as a paleoredox proxy has required recent revision. However, the basic idea behind using Cr isotopes as redox tracers is straightforward – the largest isotope fractionations are redox-dependent and occur during partial reduction of Cr(VI). As such, Cr isotopic signatures can provide novel insights into Cr redox cycling in both marine and terrestrial settings. Critically, the Cr isotope system – unlike many other trace metal proxies – can respond to short-term redox perturbations (e.g., on timescales characteristic of Pleistocene glacial-interglacial cycles). The Cr isotope system can also be used to probe the Earth's long-term atmospheric oxygenation, pointing towards low but likely dynamic oxygen levels for the majority of Earth's history.

**Keywords:** chromium isotopes, redox, oxygenation, deoxygenation, metal mass balance

ISBNs: 9781108792578 (PB), 9781108870443 (OC)
ISSNs: 2515-7027 (online), 2515-6454 (print)

# Contents

# 1 Introduction

There has been significant interest in recent decades in the use of transition metal isotope techniques to track the biogeochemical evolution of Earth's ocean-atmosphere system. In particular, transition metals, with measurable stable isotopic variability that can exist at a variety of redox states under Earth surface conditions (such as Cr, Fe, Mo, Tl, and U) have garnered much interest as potential tracers of the redox state of Earth's surface through geological time. This work has been driven by fundamental unresolved questions about past fluctuations in surface oxygen levels. For instance, surface warming over the next millennium is predicted to result in decreased levels of dissolved oxygen in seawater, potentially dramatically altering global biogeochemical cycles and reducing biological productivity and diversity in the world's oceans (Keeling et al., 2010). However, the extent of ocean deoxygenation during warming is still poorly constrained. Few studies have attempted to quantify the spatial extent of low oxygen and fully anoxic marine conditions on a global scale during past warming events. This has challenged efforts to gauge both the scope of redox-dependent feedbacks during various climatic perturbations, and the extent of future redox shifts. Equally, there is also intense debate about the magnitude and timing of larger redox shifts in deep time, including during all the major mass extinctions in Earth's history. Chromium isotopes are primed to become part of the standard toolkit – coupled with trace metal enrichments, nitrogen (N) isotopes, and other novel metal isotope systems – that we use to push forward our understanding of ocean deoxygenation.

There has also been persistent debate about Earth's long-term oxygenation over the past few decades (Lyons et al., 2014). This debate translates into significant uncertainty in the relative roles that environmental and biological factors have played in driving broad-scale evolutionary trends (Cole et al., 2020). For large, multicellular organisms like animals most of the debate has centered on whether oxygen levels were low enough to have prevented animals from establishing stable populations over million-year time scales (Butterfield, 2009; Erwin et al., 2011; Sperling et al., 2013; Towe, 1970). Quantitative constraints on oxygen levels during the majority of the Proterozoic – that is, the billion-year interval preceding the rise of animals, have been a key piece of information missing from this conversation (Kump, 2008; Lyons et al., 2014). Metal isotope systems – including Cr isotopes – can add to this debate by providing minimum or maximum constraints on oxygen concentrations at Earth's surface. Given that Cr redox cycling induces the largest fractionations in this isotope system, the extent of sedimentary Cr isotopic variability can be used to pinpoint when Cr oxidation "turned on" – locally or globally. The initiation of widespread Cr oxidation can, in principle, be quantitatively linked to a range of minimum atmospheric oxygen levels. The Cr isotope system is thus well-suited to answer questions about Earth's long-term evolution as well as recent oxygen dynamics in Earth's oceans.

Here we outline the basics of the Cr isotope system and the major remaining gaps in our knowledge of how this system works. We explore a few examples of how the Cr isotope system can provide unique insights into past fluctuations in Earth's oxygen levels. Rather than an exhaustive but cursory review of all the Cr isotope work from the past decade, we highlight a few case studies that illustrate how the Cr isotope system is well-suited to address questions regarding recent marine oxygen dynamics, as well as Earth's long-term oxygenation. We also highlight some pitfalls of previous Cr isotope work as a means of shaping future endeavors. The overall message is that although the Cr isotope system is much more complicated than originally envisioned, when applied thoughtfully, Cr isotopes remain an important part of the toolkit being used to track marine and atmospheric oxygen levels.

## 2 Basics of Cr Speciation and Isotope Fractionations

On the modern Earth's surface, the Cr cycle is largely governed by redox reactions whereby soluble Cr(VI) species are produced via oxidation of reduced Cr(III) species. Chromium is present almost exclusively in rock-forming minerals in igneous rocks containing reduced Cr(III) (Fandeur et al., 2009). Therefore, mobilization of Cr in soil environments necessitates the oxidation of Cr(III) to Cr(VI). However, because the kinetics of Cr(III) oxidation with $O_2$ are slow (Eary and Rai, 1987; Johnson and Xyla, 1991), Mn(III, IV) (oxyhydr-) oxides are typically considered to be the only environmentally relevant oxidant for Cr(III) at the Earth's surface (Bartlett and James, 1979; Eary and Rai, 1987; Fendorf, 1995). The role of superoxide, which has recently been highlighted as an important oxidant, for other metals has not been thoroughly investigated. Oxidation proceeds through dissolved Cr(III) reaction with solid phase Mn oxides and forms tetrahedrally coordinated oxyanions (e.g., $CrO_4^{2-}$, $HCrO_4^-$, $Cr_2O_7^{2-}$), which are highly soluble and readily transported in oxidizing aqueous fluids. Riverine chromate ($CrO_4^{2-}$) is thus considered to be the main source of Cr to the modern oceans (Bartlett and James, 1979; Fendorf, 1995; Konhauser et al., 2011; Oze et al., 2007). However, many rivers also contain a significant Cr(III) load (e.g., D'Arcy et al., 2016; Wu et al., 2017), which is likely bound by organic ligands.

In modern oxygenated oceans, Cr is thermodynamically stable and present predominantly as Cr(VI), although significant portions of Cr(III) exist in regions; for example, in some North Pacific water masses (Janssen et al., 2020; Wang et al., 2019). The generally conservative behavior of Cr in oxic waters is in strong contrast to that observed in anoxic systems. In anoxic systems Cr(VI) is quickly reduced by Fe(II), sulfide, solid-phase reduced Fe(II) and S phases and even some organic compounds when these are present at high concentrations (Eary and Rai, 1989; Fendorf, 1995; Patterson et al., 1997; Richard and Bourg, 1991). On reduction at circumneutral or alkaline pH, the majority of resulting Cr(III) will hydrolyze to form $Cr(OH)_3$, which is sparingly soluble and readily removed from solution. Therefore, within any anoxic aquatic system, the Cr reservoir will be largely reduced to Cr(III) and subsequently preferentially partitioned into solid (particulate) phases.

Chromium isotope compositions are reported in delta notation ($\delta^{53/52}Cr$), relative to the international standard NIST SRM-979 ($\delta^{53}Cr = 1000$ x [$(^{53}Cr/^{52}Cr)_{sample}/(^{53}Cr/^{52}Cr)_{SRM-979} - 1$]). Early work on the Cr isotope system in Cr(VI)-contaminated groundwaters established the view that surface water $\delta^{53}Cr$ values were controlled by redox transformations (Johnson and Bullen, 2004). Both theoretical and experimental studies indicate that Cr will undergo limited fractionation during most redox-independent transformations, for example, adsorption processes (Ellis et al., 2004; Johnson and Bullen, 2004; Schauble et al., 2004). The typically limited extent of redox-independent Cr isotope fractionations is largely linked to bonding preferences. Chromium(III) has a very strong preference for octahedral coordination, while Cr(VI) strongly favors a tetrahedral coordination. This is in contrast to other heavy metal isotope systems where, in a given redox state, both tetrahedral and octahedral coordination environments are common (e.g., Fe, Cu). As a result, non-redox-linked coordination changes are less likely to drive isotope fractionations of Cr than in "typical" heavy metal isotope systems (Schauble et al., 2004). Chromium(III) complexation with ligands is the most notable exception to this rule (e.g., Babechuk et al. (2018); Saad et al. (2017)). For instance, chromium(III)-organic ligand complexation can cause large (> 1‰) fractionations, but the exact mechanism driving this fractionation is currently unresolved. Saad et al. (2017) suggested that the isotope fractionation occurs during a back reaction, but additional experimental work and ab initio calculations are needed for a more detailed mechanistic understanding of most redox-independent Cr isotope fractionations (see Babechuk et al., 2018).

In marked contrast to most non-redox processes, the oxidation and reduction of Cr species induce large isotope fractionations. Because there is a narrow range of $\delta^{53}Cr$ values observed in igneous systems, with an average value of –0.124 ±0.101‰ (2SD) (Schoenberg et al., 2008), Cr isotope fractionations are often compared to this crustal range, which is canonically referred to as the igneous silicate Earth (ISE) composition. At equilibrium, the $Cr(VI)O_4^{2-}$ anion is predicted to be enriched in heavy Cr ($^{53}Cr$) by over 6‰ relative to the coexisting Cr(III) reservoir (Schauble et al., 2004). However, in natural Earth surface systems it is unlikely that the full equilibrium isotope effect will be expressed. Fractionations resulting from oxidation in natural systems observed thus far are less than 1‰. In contrast, fractionations observed during reduction from Cr(VI) to Cr(III) range between 3‰ and 5.5‰ (Ellis et al., 2002, 2004; Johnson and Bullen, 2004; Schauble et al., 2004; Zink et al., 2010). However, if reduction is quantitative, as would be expected in strongly reducing environments, large fractionations will not be expressed at the system scale (Reinhard et al., 2013).

## 3 A Global Cr Isotope Mass Balance?

### 3.1 Marine Cr Input Fluxes

The discharge-weighted riverine input of dissolved Cr to the ocean was estimated by (Reinhard et al., 2013) to be ~6 x $10^8$ mol y$^{-1}$. This estimate is significantly below that of McClain and Maher (2016), who estimated a Cr input flux roughly a factor of two higher (~1.7 x $10^9$ mol y$^{-1}$). However, recent work on catchments with minimal anthropogenic influence revealed anomalously low Cr concentrations (~4 x $10^4$ mol y$^{-1}$) relative to those from similar climate zones in the McClain and Maher (2016) study (roughly an order of magnitude lower Cr concentrations (Wu et al., 2017)). Significant anthropogenic riverine Cr contamination seems likely, with the result that accurately reconstructing the pre-anthropogenic riverine Cr flux is challenging. The potential for anthropogenic Cr contamination in rivers also affects our ability to estimate average pre-anthropogenic riverine $\delta^{53}Cr$ values. Nonetheless, it is interesting that riverine $\delta^{53}Cr$ values are highly variable, but mostly enriched towards heavy compositions relative to crustal values. Most recently, the average riverine $\delta^{53}Cr$ value was estimated to be roughly +0.47 ± 0.39‰ (Toma et al., 2019).

Hydrothermal systems do not appear to be a large Cr source, in contrast to many transition metals. High-temperature hydrothermal fluids may be depleted in Cr relative to seawater due to early mixing of hydrothermal fluids and seawater and the rapid formation of Fe (oxyhydr-)oxides, which remove Cr through co-precipitation and scavenging (German et al., 1991). Furthermore, fluid concentration anomalies combined with estimates of global heat flux associated with axial hydrothermal activity (Elderfield and Schultz, 1996) indicate that these systems amount to a small net sink of Cr from seawater (Reinhard et al., 2013). This is corroborated by low Cr inventories in sediments underlying hydrothermal systems (Bauer et al., 2019). The effect of diffuse-flow hydrothermal systems is poorly constrained, although there is some evidence that low-temperature hydrothermal fluids are only mildly enriched in Cr relative to seawater (Sander and Koschinsky, 2000). However, if these concentration anomalies are extrapolated globally by assuming the entire riverine $Mg^{2+}$ flux is removed in diffuse-flow systems, the estimated Cr flux would be ~3–4 x $10^6$ mol $y^{-1}$, still less than ~1% of the dissolved riverine flux (Reinhard et al., 2013). We also note that measurements to date indicate that the $\delta^{53}Cr$ of hydrothermal fluids is close to the ISE value (Bonnand et al., 2013).

## 3.2 Marine Cr Burial Fluxes and Isotope Mass Balance?

There are four primary marine Cr removal fluxes (Figure 1): (1) burial in sediments deposited within anoxic water columns; (2) burial in reducing continental margin sediments; (3) burial in carbonate depositional environments; and (4) burial in oxic marine sediments. Removal of Cr from seawater in any of

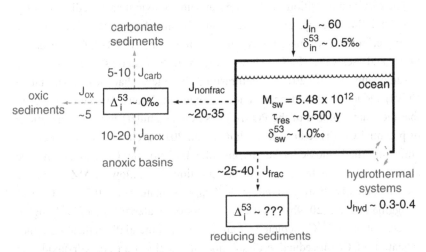

**Figure 1** Schematic of the provisional global Cr mass balance discussed in the text and revised from (Reinhard et al., 2013). The magnitude and isotopic composition of the different Cr fluxes have been updated based on Cr studies conducted in the modern oceans (Gueguen et al., 2016; Sander and Koschinsky, 2000; Sander et al., 2003; Scheiderich et al., 2015; Toma et al., 2019). Seawater Cr reservoir mass ($M_{sw}$), oceanic residence time ($\tau_{res}$), and isotope composition ($\delta^{53}_{sw}$) are shown, as are the isotopic offsets from seawater associated with burial in each sink ($\Delta^{53}_i$ terms). Numbers associated with each flux ($J_i$) are in units of $10^7$ mol $y^{-1}$. Note that the separation of sink terms into "fractionating" ($J_{frac}$) and "non-fractionating" ($J_{nonfrac}$) is provisional

these environments could, in principle, be accompanied by an isotopic fractionation – causing seawater $\delta^{53}$Cr values to deviate from the mean time-integrated input value.

### 3.2.1 Anoxic Sediments

Chromium(VI) may be efficiently scavenged from anoxic water columns, and thus Cr removal in anoxic waters likely represents a significant removal flux despite the small spatial extent of these environments. Work on sediments deposited in the Cariaco Basin, a permanently anoxic marine basin off the coast of Venezuela (Gueguen et al., 2016; Reinhard et al., 2013), suggests that authigenic Cr in euxinic (anoxic, $H_2S$-rich) sediments is isotopically similar to contemporaneous seawater $\delta^{53}$Cr values (Figure 2). The $\delta^{53}$Cr values are between 0.6‰–0.7‰, which is within error of adjacent Atlantic seawater values (Bonnand et al., 2013). The authigenic Cr isotope composition of Cariaco Basin sediments determined using a detrital element correction (using a Cr/Ti ratio) and a set of weak acid leaches are comparable (Figure 2).

The $\delta^{53}$Cr composition of Cr buried in some anoxic marine sediments may also be influenced by, and record, regional and diagenetic processes. For example, in the Peru margin oxygen minimum zone (POMZ), Cr concentrations and seawater $\delta^{53}$Cr show little variation across the redoxcline (Bruggmann et al., 2019b), contrasting the view that near quantitative Cr(VI) reduction should occur under anoxic conditions. We note, however, that the water column of the POMZ does not accumulate high concentrations of potential Cr(VI) reductants (Scholz et al., 2014), relative to other permanently stratified anoxic ocean basins like the euxinic Cariaco Basin (see above). Bulk and authigenic $\delta^{53}$Cr compositions in shallow POMZ sediments are isotopically heavy (average $\delta^{53}$Cr$_{Bulk}$ signature of 0.77 ± 0.19‰; Bruggmann et al. (2019b)). In the anoxic bottom waters of the POMZ, higher concentrations of $^{53}$Cr-depleted Cr are observed, potentially indicating remobilization of Cr on reductive dissolution of reactive Fe (oxyhydr-)oxides in surface sediments. Bruggmann et al. (2019b) also observe covariation between $\delta^{53}$Cr and organic matter contents in the POMZ sediments, which may be indicative of a small yet important authigenic Cr pool delivered to the sediment surface associated with plankton biomass. Furthermore, this organic- and Cr-rich end member has a heavy Cr isotopic composition (1.10 ± 0.08‰). Chromium isotope data from the POMZ thus reveals that even under prevailingly anoxic water column conditions, regional factors such as organic matter loading and the delivery and speciation of Fe may give rise to $\delta^{53}$Cr variations recorded in the sedimentary record.

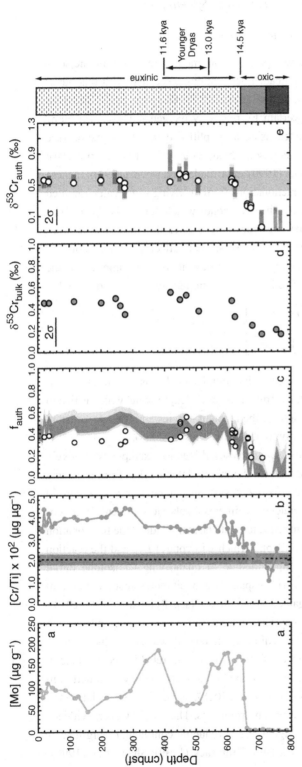

**Figure 2** Geochemical data for sediments deposited at ODP Site 1002 since the Last Glacial Maximum (reproduced from Reinhard et al., 2014). Core stratigraphy at right and data for Mo in (**a**) are from Lyons et al. (2003). Gray field in (**b**) denotes the range of Cr/Ti values for upper continental crust (UCC), while the black dotted line and red field denote the mean and 95% confidence interval for sediments deposited under oxic conditions. Values for the authigenic Cr fraction ($f_{auth}$; **c**) are calculated based on measured [Cr/Ti] values and assuming the resampled mean and 95% confidence interval of oxic sediments as the detrital background (red field) or the range of estimates for UCC (gray field: Rudnick 2003). In (**c**), are values calculated based on leachate [Cr] data (open circles). The Cr isotope composition of bulk sediments is shown in (**d**). In (**e**), the isotopic composition of the reconstructed authigenic component ($\delta^{53}Cr_{auth}$) is shown, according to calculations using the UCC composite range (gray bars), the mean and 95% confidence interval for oxic sediments (red bars), and the sediment leaches (open circles). The blue shaded region denotes the range for modern deep Atlantic seawater (Bonnand et al., 2013). Error bars in (**d**, **e**) show external reproducibility ($2\sigma$) of ±0.1‰

### 3.2.2 Reducing Marine Sediments

There is limited published Cr isotope data from modern reducing marine sediments, however, there are significant fractionations associated with this burial term (Bauer et al., 2018; Bruggmann et al., 2019b; Gueguen et al., 2016). In reducing marine sediments overlain by relatively oxygenated waters, isotope fractionations are often predicted using a simplified diagenetic equation (see Bauer et al. (2018); Clark and Johnson (2008); Johnson and DePaolo (1994); following the approach of Bender (1990) and Brandes and Devol (1997)). In these sedimentary systems, isotope fractionations from overlying water values are tied to the scale of the nonreactive diffusive zone, which for Cr is the oxygen penetration depth. The thinner this zone, the greater the fractionation from overlying water. The rate constant will also strongly affect the fractionation by controlling how close the system is to near quantitative consumption of the porewater Cr reservoir. The "effective" fractionation factor ($^{53}\alpha_{eff}$) is defined by:

$$^{53}\alpha_{eff} = \sqrt{^{53}\alpha_{int}} \left[ \frac{1 + \left( \frac{L_{diff}}{\lambda} \right)}{1 + \left( \frac{L_{diff}}{\lambda} \right) \sqrt{^{53}\alpha_{int}}} \right]$$

where $^{53}\alpha_{int}$ is the intrinsic isotope fractionation factor; $L_{diff}$ is the length scale of the nonreactive zone and $\lambda$ is the diffusion-reaction length scale, which is tied to the sediment diffusion coefficient of the reactant species and the rate constant for a first-order reduction reaction (Clark and Johnson, 2008).

As the diffusive length scale ($L_{diff}$) increases, driven for example by relatively deep $O_2$ penetration into the sediment column, the isotopic offset from overlying water decreases, and for strongly oxidizing systems will be near zero (Figure 3) (Clark and Johnson, 2008). Reducing sediments ("suboxic sediments") in contrast should be fractionated from seawater with the magnitude of the fractionation scaling with oxygen penetration depth, which will control $L_{diff}$ and the reaction rate term. In principle, this simple method of quantifying Cr isotope burial fractionations could be coupled with a spatially explicit representation of bottom water oxygen levels and organic carbon fluxes (see Reinhard et al., 2013). Additionally, this framework may lead to a proxy for local oxygen penetration depths in past reducing marine sediments (if coeval seawater values can be independently constrained and authigenic Cr can be reliably reconstructed).

The idealized 1D model has been applied effectively to Cr(VI) reduction in lacustrine-reducing sediments (Bauer et al., 2018), however, its scope has yet to be fully explored in modern marine environments. There is, however, evidence for elevated burial rates of authigenic Cr in reducing continental margin sediments (e.g., Bruggmann et al. (2019b); Brumsack (1989); Shaw et al. (1990))

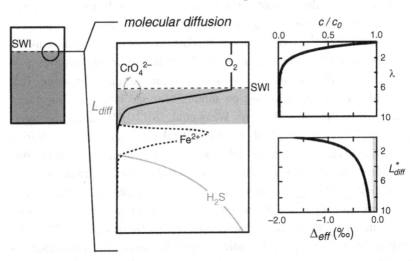

**Figure 3** Idealized reaction-transport scenario for Cr redox cycling in reducing marine sediments (after Reinhard et al., 2014). In a reducing sediment system, transport of $CrO_4^{2-}$ occurs through diffusion. A zone may exist between the sediment-water interface (SWI) and a deeper anoxic zone where reduction takes place. Through this zone, $CrO_4^{2-}$ diffuses but does not react. The length scale of this nonreactive zone is given by $L_{diff}$ (blue bar). Shown at right are (top) normalized $CrO_4^{2-}$ concentration ($c/c_0$) as a function of $e$-folding depth below the top of the reaction zone (in units of $\lambda = \sqrt{D_{Cr}/k}$, where $D_{Cr}$ is the diffusion coefficient and $k$ is the first-order reaction rate constant) and (bottom) isotopic offset from overlying water ($\Delta_{eff}$) as a function of dimensionless depth ($L_{diff}^{*}$ $= L_{diff}/\lambda$; see text)

and although these environments are restricted spatially, they likely account for a large portion of the net removal flux of Cr from the ocean. This is important, as we expect that these settings will most readily express the isotopic effects of partial reduction of seawater $CrO_4^{2-}$. As a result, the mass-weighted isotopic burial flux from the ocean in such regions might be expected to respond dynamically to changes in bottom water chemistry along continental margins – and thus ocean oxygen levels.

### 3.2.3 Carbonate Sediments

There is currently little Cr isotope data from modern marine carbonate sediments. Sparse results suggest that Bahamian ooids, which typically give a mean radiocarbon age of a few thousand years, match Atlantic dissolved seawater $\delta^{53}Cr$ values (~0.6‰) (Bonnand et al., 2013). It may be expected that carbonate sediments would be minimally fractionated from contemporaneous seawater $\delta^{53}Cr$ values due to the lack of coordination change as the chromate oxyanion is incorporated into the carbonate lattice. However, it is possible that there are vital effects during biogenic carbonate formation. This is consistent with observations on foraminifera and bivalves (Bruggmann et al., 2019a; Pereira et al., 2016; Wang et al., 2017). Chromium concentration and $\delta^{53}Cr$ data obtained in foraminiferal shells from modern marine sediments, water columns, and laboratory cultures, furthermore, imply that Cr uptake into the carbonate lattice occurs post-depositionally (Remmelzwaal et al., 2019). This opens the possibility that Cr signatures in foraminiferal shells record bottom and porewater signals, which is contrary to interpretations that these shells are a reliable archive of surface seawater $\delta^{53}Cr$ in the geologic past. Although some studies suggest diagenetic effects on carbonate-bound Cr isotope systematics may be important, the precise mechanisms remain poorly constrained. Working out these details in future studies will be of value for the Cr isotope paleoredox proxy.

### 3.2.4 Oxic Marine Sediments

Currently published data from oxic marine sediments is also limited (Bruggmann et al., 2019b; Gueguen et al., 2016; Wei et al., 2018). A reasonable null hypothesis is that Cr removal in such environments will occur without significant isotopic offset from contemporaneous seawater – given that there is no observed fractionation during Cr sorption to metal (Fe, Al) oxides (Ellis et al., 2004), which is likely the main Cr removal pathway in oxic marine sediments. In addition, sorption efficiency of $CrO_4^{2-}$ on these phases is minimal at typical marine pH values (discussed above), providing an important and potentially very useful contrast to transition metal isotope systems such as molybdenum (Mo). However, $\delta^{53}Cr$ values in Fe-Mn crusts are highly variable, likely pointing towards Fe-Mn-induced Cr redox cycling (Wei et al., 2018).

## 3.3 A Framework for Marine Cr Isotope Mass Balance

Although the material fluxes in the Earth surface Cr cycle can be roughly estimated, the isotopic fluxes are poorly constrained at present. Nevertheless, the mechanisms that remove Cr from the ocean will either capture the isotopic composition of ambient source waters or they will be isotopically offset – with more negative $\delta^{53}Cr$ values than seawater. Isotope mass balance for Cr within the ocean can thus be generally formulated according to:

$$\frac{d\delta_{sw}^{53}}{dt} = \left[ J_{in}\left(\delta_{in}^{53} - \delta_{sw}^{53}\right) - \sum_i J_i \Delta_i^{53} \right] M_{sw}^{-1}$$

where $J_{in}$ denotes the combined input flux integrates across all sources, $\delta_{sw}^{53}$ and $\delta_{in}^{53}$ denote the $\delta^{53}Cr$ composition of seawater (in a globally integrated sense) and the $\delta^{53}Cr$ composition of combined inputs to the ocean, respectively, and $M_{sw}$ denotes the mass of Cr in seawater. The last term in brackets represents the combined effects of removal fluxes ($J_i$) that are isotopically offset from seawater by a factor $\Delta_i^{53}$ that depends on the governing removal process. If we assume steady state ($d^{\delta_{sw}^{53}}/dt = 0$), and specify $f_i = J_i/J_{in}$, we arrive at a conventional steady state expression for oceanic Cr isotope mass balance:

$$\delta_{sw}^{53} = \delta_{in}^{53} - \sum_i J_i \Delta_i^{53}$$

Chromium removal fluxes that are isotopically similar or identical to seawater do not explicitly appear in the isotope mass balance expression, but are important through their role in controlling the value of $f_i$. This approach applies to an idealized "mean" Cr isotope value of seawater, which is difficult to precisely determine as there is significant marine Cr isotope variability (see below) (Scheiderich et al., 2015; Wang et al., 2019). Nonetheless, the approach provides a meaningful framework for exploring the controls on $\delta_{sw}^{53}$ values. Inspection of the above equations in the context of our provisional mass balance framework shows that seawater $\delta^{53}Cr$ values will either directly track the time-integrated mean input value or will be offset as a result of isotopic fractionations imposed during removal into reducing marine sediments. Indeed, some $\delta^{53}Cr$ values recorded in modern marine sediments do indicate offset from contemporaneous seawater, implying complex local controls on $\delta^{53}Cr$ compositions in these environments (Bruggmann et al., 2019b) (see sections above). Therefore, as additional $\delta^{53}Cr$ values from the modern oceans become available, these new constraints may be used to revise the current Cr mass balance model.

## 4 Seawater Cr Isotope Values

The $\delta^{53}Cr$ values measured in seawater are relatively $^{53}Cr$-enriched, falling within a range of $\sim 0.25$ to $1.75‰$ (Bonnand et al., 2013; Goring-Harford et al., 2018; Janssen et al., 2020; Moos and Boyle, 2019; Paulukat et al., 2016; Rickli et al., 2019; Scheiderich et al., 2015). Coupled seawater [Cr] and $\delta^{53}Cr$ measurements imply that most seawater $\delta^{53}Cr$ values in the global oceans can be described using a Rayleigh fractionation law, operating with an isotopic fractionation factor of $\varepsilon \approx 0.85‰$ (Figure 4) (Scheiderich et al., 2015). The underlying mechanisms that produce this fractionation factor are poorly constrained, however, it likely results from a combination of processes such as: Cr(VI) reduction to Cr(III) in oxygen minimum zones (Moos et al., 2020; Wang et al., 2019), biological productivity and the removal of isotopically light Cr(III) associated with organic matter (Goring-Harford et al., 2018; Janssen et al., 2020), and mixing of isotopically distinct water masses (Rickli et al., 2019). Ancient and more recent sediments deposited on the seafloor have been commonly proposed to capture ambient seawater $\delta^{53}Cr$ values (e.g., Crowe et al. (2013); Frei et al. (2009); Gueguen et al. (2016); Reinhard et al. (2014)). However, few marine sediment samples appear to record contemporaneous modern seawater $\delta^{53}Cr$ values, making it difficult to surmise under what conditions the sedimentary reservoir faithfully captures the Cr isotope composition of seawater and whether this is altered depending on lithology or subsequent diagenesis (Bruggmann et al., 2019b; Goring-Harford et al., 2018). More recent seawater $\delta^{53}Cr$ measurements also reveal numerous ocean sites that do not plot on the Rayleigh-type fractionation line (Paulukat et al., 2016; Wang et al., 2019), implying some heterogeneity with respect to Cr concentration and isotopic composition. As precise links between the $\delta^{53}Cr$ composition of seawater, water column redox state, and biological productivity remain unclear, more Cr isotope data is required to better constrain local vs. global scale processes within the marine Cr cycle and their associated implications for the $\delta^{53}Cr$ seawater composition of the global oceans.

## 5 The Sedimentary Cr Isotope Record

### 5.1 The Early Cr Record

Much of the deep time Cr isotope record so far has focused on searching for early evidence of oxidative Cr cycling. In a landmark paper, Frei et al. (2009) reported Cr isotope data from iron formations ranging in age from ~3.8 Ga to 0.635 Ga, which they used to examine the progressive oxygenation of Earth's atmosphere. Iron formations were proposed to be ideal sedimentary Cr isotope archives, given that these rocks have a small detrital Cr component and that the initial iron phases trap large amounts of authigenic Cr without a substantial isotopic fractionation (Frei et al., 2009). This initial survey indicated that there is a pronounced difference in the Cr isotope composition of Archean and Paleoproterozoic iron formations compared to those of Neoproterozoic age. Older iron formations deposited between ~3.8 and 1.8 Ga have Cr isotope values (0.28‰ to –0.28‰) that fall almost entirely within two standard deviations of a mean of –0.09‰ and thus very close to

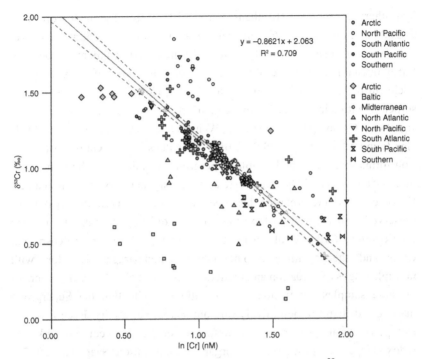

**Figure 4** Cross plot of logarithmic Cr concentration and $\delta^{53}$Cr data for seawater samples (n =321). The solid line is a linear regression using data points obtained at waters depths > 50 m (colored symbols). The dashed lines represent a 95% confidence interval of the mean. Gray symbols represent data points obtained at water depths < 50 m. Data from the Arctic (Paulukat et al., 2016; Scheiderich et al., 2015). Data from the Baltic (Paulukat et al., 2016). Data from the Mediterranean (Paulukat et al., 2016a). Data from the North Atlantic (Bonnand et al., 2013; Paulukat et al., 2016; Sun et al., 2019). Data from the North Pacific (Janssen et al., 2020; Moos and Boyle, 2019; Moos et al., 2020; Scheiderich et al., 2015; Wang et al., 2019). Data from the South Atlantic (Bonnand et al., 2013; Goring-Harford et al., 2018; Holmden et al., 2016; Pereira et al., 2015). Data from the South Pacific (Bruggmann et al., 2019b; Farkas et al., 2018; Moos and Boyle, 2019; Moos et al., 2020; Paulukat et al., 2016). Data from the Southern Ocean (Paulukat et al., 2016; Rickli et al., 2019).

the isotopic value of their igneous sources (Schoenberg et al., 2008). Some of the small (up to 0.28‰) positive Cr isotope fractionations preserved in Archean and Paleoproterozoic iron formations were proposed to record a signal for early oxidative processes (Frei et al., 2009), although it has also been suggested that small positive fractionations could also be linked to hydrothermal processes (Konhauser et al., 2011). In subsequent work, Frei and others (Frei et al., 2016) found significant Cr isotope fractionations in ca. 3.8 Ga iron formations from Isua, some of the oldest sedimentary rocks on Earth. Intriguingly, these results suggest a very early onset of an oxidative Cr cycle (in the framework of Frei et al., 2009). However, there is extensive serpentinization in this succession, and this process has been shown to induce significant Cr isotope fractionations (Wang et al., 2016a). Crowe et al. (2013) also found a large range (>0.8‰) of $\delta^{53}$Cr values in 2.95 Ga paleosols and iron formations from the Pongola Supergroup, which, given the framework outlined above, should be interpreted as a clear signal for oxidative processes. However, this interpretation has been questioned, given that two well-preserved drill cores from the same interval did not show similar isotope fractionations (Albut et al., 2018; Colwyn et al., 2019). The paleosol that Crowe and others examined also has a complicated paragenetic history, with multiple stages of oxidation and reduction (Crowe et al., 2013). However, all of these samples are from different regions of the Pongola Supergroup making it difficult to definitively rule out spatial variability. Even if the Cr isotope values are primary, it is important to note that they could be linked to redox independent processes (e.g., organic complexation (Saad et al., 2017)). Intriguingly, new work presents evidence for photosynthetic biomineraliza-tion of manganese oxides – a key mineral catalyst required for Cr(III) oxida-tion – in the absence of molecular $O_2$ (Daye et al., 2019). Indeed, if such a metabolism were widespread on early Earth, this would have implications for interpreting Cr isotope fractionations and links to atmospheric oxygen. In sum, there are hints of a very early onset of Cr oxidation at Earth's surface, but definitive evidence for this process remains elusive.

## 5.2 The Proterozoic Cr Record

Initial Cr isotope work in Proterozoic successions was focused on nearshore iron-rich marine units. This work built directly from earlier Cr isotope work on Archean and Neoproterozoic iron formations. Nearshore deposits were targeted to avoid any obvious contribution from contemporaneous deeper hydrothermal systems (Planavsky et al., 2014). The targeted granular deposits are sedimento-logically and geochemically equivalent to Phanerozoic nearshore iron-rich deposits – rocks commonly referred to as ironstones – and formed in marginal marine or deltaic settings genetically similar to iron-rich portions of the modern Amazon delta (Aller et al., 1986). The examined mid-Proterozoic ironstones have igneous (unfractionated) $\delta^{53}$Cr values (–0.124‰, ±0.101 2SD; Figure 5). This is in marked contrast to Phanerozoic ironstones, which have consistently

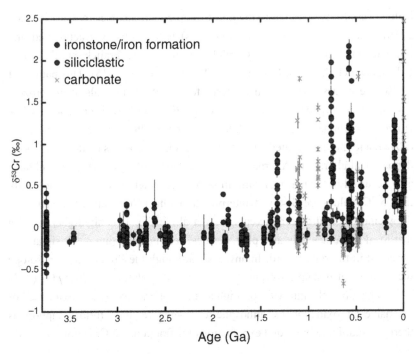

**Figure 5** The current sedimentary Cr isotope record. Persistently positive Cr isotope values appear in the record in the Proterozoic. Most positive values are likely linked to Cr(III) oxidation in soils, but these values could also be tied to marine Cr(III) oxidation or organic complexation. We note that there is also potential that some positive Cr isotope values may be the result of recent weathering and alteration (see main text). Figure updated from Planavsky et al. (2018).

positive $\delta^{53}$Cr values indicative of oxidative Cr cycling. This observation is most parsimoniously explained by a switch from a largely unfractionated, Cr(III)-dominated terrestrial-marine flux to an isotopically heavy, mixed Cr(III) and Cr(VI) terrestrial-marine flux. Indeed, this interpretation is supported by lack of petrographic evidence for a detrital contribution (which would likely have an unfractionated isotopic composition) to these sediments (Cole et al., 2018). This data suggests there were at least intervals of mid-Proterozoic time during which atmospheric oxygen was less than ~1% PAL – assuming that quantitative iron oxidation in the soil realm is needed to prevent pervasive Cr reduction (Planavsky, 2018).

There has been extensive subsequent mid-Proterozoic Cr isotope work in a range of different sedimentary archives, with units characterized by limited Cr isotope variability – similar to the mid-Proterozoic ironstones (e.g., Cole et al., 2016) – and units (particularly carbonates) with a mix of fractionated and unfractionated $\delta^{53}Cr$ values (Canfield et al., 2018; Fralick et al., 2017; Gilleaudeau et al., 2016; Rodler et al., 2016; Sial et al., 2015). This pattern of Cr isotope values was originally interpreted to reflect intervals of Proterozoic atmospheric oxygenation (Canfield et al., 2018; Gilleaudeau et al., 2016). However, carbonates are also extremely prone to alteration (e.g., Hood et al. (2018)) and the fractionated Cr isotope values in carbonates could reflect post-depositional alteration. Alternatively, some degree of Cr oxidation could have also occurred in the marine realm. In some Neoproterozoic rocks, isotopically light $\delta^{53}Cr$ values are observed, interpreted to reflect Cr isotope fractionation in the vicinity of hydrothermal vents (Sial et al., 2015). This highlights the idea that local processes may also play a critical role in imparting $\delta^{53}Cr$ signatures observed in the rock record. In any case, although the Proterozoic Cr isotope record hints at dynamic oxygenation – with intervals of very low $pO_2$ (<1% PAL) and possible intervals of higher atmospheric oxygen – evidence for extensive Cr oxidation by atmospheric oxygen requires further support as there are multiple nonunique explanations for fractionated Cr isotope values.

## 5.3 The Phanerozoic Record

There have been several studies of Cr isotope behavior in the Phanerozoic, through major biogeochemical perturbations and periods of biogeochemical stasis. Overall, there is a predominance of positive authigenic Cr isotope values in Phanerozoic sediments. This is consistent with the view that seawater through the Phanerozoic has been isotopically heavy (typically between 0.5 and 1.5‰; see most recent compilation in Planavsky (2018)) and that the detrital Cr reservoir (which is larger than the authigenic Cr reservoir (Meybeck, 1994)) has retained slightly subcrustal Cr isotope values. The observed range of authigenic values (~−1.0‰ to +2.0‰) also suggests that the full potential equilibrium fractionation during Cr(VI) reduction (>5‰) is not expressed during Cr removal from the oceans (Bruggmann et al., 2019b; Cole et al., 2016; Gueguen et al., 2016; Holmden et al., 2016; Reinhard et al., 2014; Wang et al., 2016b).

There are several documented short-term perturbations in seawater Cr isotope values. For instance, in several locations, there are coherent decreases in Cr isotope values in anoxic shales deposited during the Cretaceous Ocean Anoxic Event 2 (OAE2) (Holmden et al., 2016; Wang et al., 2016b). The most straightforward interpretation is that these records capture evidence of an expansion of anoxic sediments at the expense of suboxic sediments (Wang et al., 2016b). This model builds on the observation that, in euxinic settings, there is near quantitative capture of Cr, resulting in the absence of an isotopic reservoir effect and a record of seawater Cr isotope values (Gueguen et al., 2016). This is in contrast to reducing sediments overlain by oxic waters that are likely to induce significant isotopic fractionations (Bauer et al. (2018); Bruggmann et al. (2019b); Gilleaudeau et al. (2016); Reinhard et al. (2014); see above). These results reveal that Cr isotopes can provide unique insight – relative to myriad other redox proxies that have been studied across OAE2 – as a signal for a shift in the ratio of anoxic to suboxic marine settings. There are also coherent Cr isotope shifts on glacial-interglacial timescales in a core from the Peru Margin (Hole 680A; Figure 6). Heavy $\delta^{53}$Cr compositions are observed during interglacial intervals, consistent with more extensive anoxic (but not sulfidic) conditions in during these warmer periods (Gueguen et al., 2016). As noted above, water column Cr reduction has recently been observed in some of these settings (e.g., Eastern Tropical Pacific OMZ) (Moos et al., 2020; Wang et al., 2019). This interpretation is bolstered by the strong correlation between sedimentary Cr and N isotope values – given the well-documented water column denitrification control on sedimentary N isotope values.

A fast-growing database of $\delta^{53}$Cr data from modern sediments and ambient waters indicates that Cr isotope fractionations are highly complex. For example, in some oxygen-depleted ocean waters, Cr removal is associated with large Cr isotope fractionations (Moos et al., 2020). Yet, in other oxygen-depleted regions of the ocean these phenomena are absent (Janssen et al., 2020), implying that the relationship between $\delta^{53}$Cr and water column oxidation state may be nuanced (Nasemann et al., 2020). Instead, local factors such as biological productivity (Janssen et al., 2020; Rickli et al., 2019) and diagenetic processes (Bruggmann et al., 2019b; Goring-Harford et al., 2018), likely play important roles. Further exploration of the Cr isotope record in modern marine environments and across perturbations throughout the Phanerozoic Eon are obvious and important areas for future research.

## 6 Summary and Future Directions

The redox state of the oceans exerts a first order control on marine ecosystem structure and regulates several key biogeochemical cycles (e.g., C, S, P). Therefore, tracking redox shifts through Earth's history is essential in determining the role of anoxia in driving biotic transitions. In this regard, the Cr isotope system may provide a powerful paleoredox proxy. Most major isotope

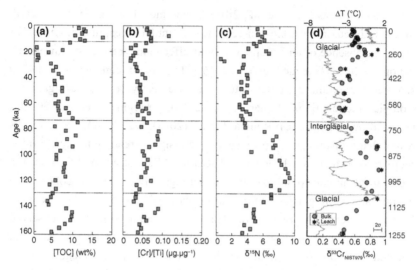

**Figure 6** Geochemical data from Peru Margin (Hole 680A) showing trends over the most recent interglacial glacial cycle. There are clear trends in Cr isotope data that track sedimentary N isotope values from the same core. These swings in Cr isotope values have been interpreted as track variation in the local (the south Pacific Eastern equatorial upwelling zone) extent of anoxia. During the interglacial interval there was more extensive anoxia at the core of the oxygen minimum zone – a setting with a limited Cr isotope fractionation – than during the glacial interval. Figure from Gueguen et al. (2016).

fractionations in the Cr isotope system are linked to redox reactions, and because Cr has a short marine residence time the system is well-suited to track redox shifts that may occur during rapid climate perturbations. Better constraints on shifts in the redox landscape that occur during rapid warming events will aid efforts to quantify the impact of warming on marine oxygen levels over the next millennium. The Cr isotope system has also been used extensively to track Earth's oxygen levels, and Cr isotopes are one of several proxies that suggest atmospheric oxygen levels were low (<1% PAL) for the majority of Earth's history. However, the utility of Cr isotope measurements is dependent on our understanding of the behavior of Cr in modern systems, which, on the whole, remains nascent. There are still many gaps in our knowledge of the principal controls on the modern global Cr isotope mass balance, and only a basic understanding of the causes of Cr isotope variations in seawater (spatially and temporally). For example, to date, very limited species-specific $\delta^{53}$Cr data exists. Studies that aim to produce species-specific $\delta^{53}$Cr data in modern environments (Wang et al., 2019), therefore, will improve our understanding of Cr isotope fractionations associated with a plethora of important processes including redox transformations. With continued advances in our calibration of the Cr isotope system, this proxy will hopefully offer a new look at the magnitude of shifts in marine anoxia, and perhaps biological productivity, during past warming events and will continue to drive refinement of quantitative constraints on atmospheric $O_2$ levels in Earth's deep past.

# References

Albut, G., Babechuk, M. G., Kleinhanns, I. C. et al., 2018, Modern rather than Mesoarchaean oxidative weathering responsible for the heavy stable Cr isotopic signatures of the 2.95 Ga old Ijzermijn iron formation (South Africa): Geochimica et Cosmochimica Acta, v. 228, pp. 157–89.

Aller, R. C., Mackin, J. E., and Cox, R. T., 1986, Diagenesis of Fe and S in Amazon Inner Shelf muds – apparent dominance of Fe reduction and implications for the genesis of ironstones: Continental Shelf Research, v. 6, no. 1–2, pp. 263–89.

Babechuk, M. G., Kleinhanns, I. C., Reitter, E., and Schoenberg, R., 2018, Kinetic stable Cr isotopic fractionation between aqueous Cr(III)-Cl-H2O complexes at 25 degrees C: Implications for Cr(III) mobility and isotopic variations in modern and ancient natural systems: Geochimica et Cosmochimica Acta, v. 222, pp. 383–405.

Bartlett, R., and James, B., 1979, Behavior of chromium in soils: III. Oxidation: J. Environ. Qual., v. 8, no. 1, pp. 31–5.

Bauer, K. W., Cole, D. B., Asael, D. et al., 2019, Chromium isotopes in marine hydrothermal sediments: Chemical Geology, p. 119286.

Bauer, K. W., Gueguen, B., Cole, D. B. et al., 2018, Chromium isotope fractionation in ferruginous sediments: Geochimica et Cosmochimica Acta, v. 223, pp. 198–215.

Bender, M. L., 1990, The $\delta^{18}O$ of dissolved $O_2$ in seawater: A unique tracer of circulation and respiration in the deep sea: Journal of Geophysical Research, v. 95, pp. 22243–52.

Bonnand, P., James, R., Parkinson, I., Connelly, D., and Fairchild, I., 2013, The chromium isotopic composition of seawater and marine carbonates: Earth and Planetary Science Letters, v. 382, pp. 10–20.

Brandes, J. A., and Devol, A. H., 1997, Isotopic fractionation of oxygen and nitrogen in coastal marine sediments: Geochimica et Cosmochimica Acta, v. 61, pp. 1793–801.

Bruggmann, S., Klaebe, R. M., Paulukat, C., and Frei, R., 2019a, Heterogeneity and incorporation of chromium isotopes in recent marine molluscs (Mytilus): Geobiology, v. 17, no. 4, pp. 417–35.

Bruggmann, S., Scholz, F., Klaebe, R., Canfield, D., and Frei, R., 2019b, Chromium isotope cycling in the water column and sediments of the Peruvian continental margin: Geochimica et Cosmochimica Acta, v. 257, pp. 224–242.

Brumsack, H. J., 1989, Geochemistry of recent TOC-rich sediments from the Gulf of California and the Black Sea: Geologische Rundschau, v. 78, pp. 851–82.

Butterfield, N. J., 2009, Oxygen, animals and oceanic ventilation: An alternative view: Geobiology, v. 7, no. 1, pp. 1–7.

Canfield, D. E., Zhang, S. C., Frank, A. B. et al., 2018, Highly fractionated chromium isotopes in Mesoproterozoic-aged shales and atmospheric oxygen: Nature Communications, v. 9, pp. 1–11.

Clark, S. K., and Johnson, T. M., 2008, Effective isotopic fractionation factors for solute removal by reactive sediments: A laboratory microcosm and slurry study: Environmental Science & Technology, v. 42, pp. 7850–55.

Cole, D. B., Mills, D. B., Erwin, D. H. et al., 2020, On the co-evolution of surface oxygen levels and animals: Geobiology, v. 18, no. 3, pp. 260–81.

Cole, D. B., O'Connell, B., and Planavsky, N. J., 2018, Authigenic chromium enrichments in Proterozoic ironstones: Sedimentary Geology, v. 372, pp. 25–43.

Cole, D. B., Reinhard, C. T., Wang, X. L. et al., 2016, A shale-hosted Cr isotope record of low atmospheric oxygen during the Proterozoic: Geology, v. 44, no. 7, pp. 555–8.

Colwyn, D. A., Sheldon, N. D., Maynard, J. B. et al., 2019, A paleosol record of the evolution of Cr redox cycling and evidence for an increase in atmospheric oxygen during the Neoproterozoic: Geobiology, v. 17, no. 6, pp. 579–93.

Crowe, S. A., Dossing, L. N., Beukes, N. J. et al., 2013, Atmospheric oxygenation three billion years ago: Nature, v. 501, no. 7468, pp. 535–538.

D'Arcy, J., Babechuk, M. G., Dossing, L. N., Gaucher, C., and Frei, R., 2016, Processes controlling the chromium isotopic composition of river water: Constraints from basaltic river catchments: Geochimica et Cosmochimica Acta, v. 186, pp. 296–315.

Daye, M., Klepac-Ceraj, V., Pajusalu, M. et al., 2019, Light-driven anaerobic microbial oxidation of manganese: Nature, v. 576, no. 7786, pp. 311–314.

Eary, L. E., and Rai, D., 1987, Kinetics of chromium(III) oxidation to chromium(VI) by reaction with manganese dioxide: Environmental Science & Technology, v. 21, no. 12, pp. 1187–93.

Eary, L. E., and Rai, D., 1989, Kinetics of chromate reduction by ferrous ions derived from hematite and biotite at 25 degrees C: American Journal of Science, v. 289, no. 2, pp. 180–213.

Elderfield, H., and Schultz, A., 1996, Mid-ocean ridge hydrothermal fluxes and the chemical composition of the ocean: Annual Review of Earth and Planetary Sciences, v. 24, pp. 191–224.

Ellis, A. S., Johnson, T. M., and Bullen, T. D., 2002, Chromium isotopes and the fate of hexavalent chromium in the environment: Science, v. 295, no. 5562, pp. 2060–62.

Ellis, A. S., Johnson, T. M., and Bullen, T. D., 2004, Using chromium stable isotope ratios to quantify Cr(VI) reduction: Lack of sorption effects: Environmental Science & Technology, v. 38, no. 13, pp. 3604–7.

Erwin, D. H., Laflamme, M., Tweedt, S. M. et al., 2011, The Cambrian conundrum: Early divergence and later ecological success in the early history of animals: Science, v. 334, no. 6059, pp. 1091–7.

Fandeur, D., Juillot, F., Morin, G. et al., 2009, Synchrotron-based speciation of chromium in an oxisol from New Caledonia: Importance of secondary Fe-oxyhydroxides: American Mineralogist, v. 94, no. 5–6, pp. 710–19.

Farkas, J., Fryda, J., Paulukat, C. et al., 2018, Chromium isotope fractionation between modern seawater and biogenic carbonates from the Great Barrier Reef, Australia: Implications for the paleo-seawater delta Cr-53 reconstruction: Earth and Planetary Science Letters, v. 498, pp. 140–51.

Fendorf, S. E., 1995, Surface reactions of chromium in soils and waters: Geoderma, v. 67, no. 1–2, pp. 55–71.

Fralick, P., Planavsky, N., Burton, J.et al. R., 2017, Geochemistry of Paleoproterozoic gunflint formation carbonate: Implications for hydrosphere-atmosphere evolution: Precambrian Research, v. 290, pp. 126–46.

Frei, R., Crowe, S. A., Bau, M. et al., 2016, Oxidative elemental cycling under the low O-2 Eoarchean atmosphere: Scientific Reports, v. 6, pp. 210–58.

Frei, R., Gaucher, C., Poulton, S. W., and Canfield, D. E., 2009, Fluctuations in Precambrian atmospheric oxygenation recorded by chromium isotopes: Nature, v. 461, pp. 250–4.

German, C. R., Campbell, A. C., and Edmond, J. M., 1991, Hydrothermal scavenging at the Mid-Atlantic Ridge: Modification of trace element dissolved fluxes: Earth and Planetary Science Letters, v. 107, pp. 101–14.

Gilleaudeau, G. J., Frei, R., Kaufman, A. J. et al., 2016, Oxygenation of the mid-Proterozoic atmosphere: Clues from chromium isotopes in carbonates: Geochemical Perspectives Letters, v. 2, no. 2, pp. 178–187.

Goring-Harford, H. J., Klar, J. K., Pearce, C. R. et al., 2018, Behaviour of chromium isotopes in the eastern sub-tropical Atlantic Oxygen Minimum Zone: Geochimica et Cosmochimica Acta, v. 236, pp. 41–59.

Gueguen, B., Reinhard, C. T., Algeo, T. J. et al., 2016, The chromium isotope composition of reducing and oxic marine sediments: Geochimica et Cosmochimica Acta, v. 184, pp. 1–19.

Holmden, C., Jacobson, A., Sageman, B., and Hurtgen, M., 2016a, Response of the Cr isotope proxy to Cretaceous Ocean Anoxic Event 2 in a pelagic

carbonate succession from the Western Interior Seaway: Geochimica et Cosmochimica Acta, v. 186, pp. 277–95.

Hood, A. V., Planavsky, N. J., Wallace, M. W., and Wang, X. L., 2018, The effects of diagenesis on geochemical paleoredox proxies in sedimentary carbonates: Geochimica et Cosmochimica Acta, v. 232, pp. 265–87.

Janssen, D. J., Rickli, J., Quay, P. D. et al., 2020, Biological control of chromium redox and stable isotope composition in the surface ocean: Global Biogeochemical Cycles, v. 34, no. 1. https://doi.org/10.1029/2019GB006397

Johnson, C. A., and Xyla, A. G., 1991, The oxidation of chromium(III) to chromium(VI) on the surface of manganite ($\gamma$-MnOOH): Geochimica et Cosmochimica Acta, v. 55, no. 10, pp. 2861–6.

Johnson, T. M., and Bullen, T. D., 2004, Mass-dependent fractionation of selenium and chromium isotopes in low-temperature environments: Reviews in Mineralogy and Geochemistry, v. 55, no. 1, pp. 289–317.

Johnson, T. M., and DePaolo, D. J., 1994, Interpretation of isotopic data in groundwater-rock systems: Model development and application to Sr isotope data from Yucca Mountain: Water Resources Research, v. 30, pp. 1571–87.

Keeling, R. F., Kortzinger, A., and Gruber, N., 2010, Ocean deoxygenation in a warming world: Annual Review of Marine Science, v. 2, pp. 199–229.

Konhauser, K. O., Lalonde, S. V., Planavsky, N. J. et al., 2011, Aerobic bacterial pyrite oxidation and acid rock drainage during the Great Oxidation Event: Nature, v. 478, no. 7369, pp. 369–373.

Kump, L., 2008, The rise of atmospheric oxygen: Nature, v. 451, pp. 277–8.

Lyons, T. W., Reinhard, C. T., and Planavsky, N. J., 2014, The rise of oxygen in Earth's early ocean and atmosphere: Nature, v. 506, no. 7488, pp. 307–15.

Lyons, T. W., Werne, J. P., Hollander, D. J., and Murray, R. W., 2003, Contrasting sulfur geochemistry and Fe/Al and Mo/Al ratios across the last oxic-to-anoxic transition in the Cariaco Basin, Venezuela: Chemical Geology, v. 195, pp. 131–57.

McClain, C. N., and Maher, K., 2016, Chromium fluxes and speciation in ultramafic catchments and global rivers: Chemical Geology, v. 426, pp. 135–57.

Moos, S. B., and Boyle, E. A., 2019, Determination of accurate and precise chromium isotope ratios in seawater samples by MC-ICP-MS illustrated by analysis of SAFe Station in the North Pacific Ocean: Chemical Geology, v. 511, pp. 481–93.

Moos, S. B., Boyle, E. A., Altabet, M. A., and Bourbonnais, A., 2020, Investigating the cycling of chromium in the oxygen deficient waters of the Eastern Tropical North Pacific Ocean and the Santa Barbara Basin using stable isotopes: Marine Chemistry, v. 221. https://doi.org/10.1016/j.marchem.2020.103756

Nasemann, P. H., Janssen, D. J., Rickli, J. et al., 2020, Chromium reduction and associated stable isotope fractionation restricted to anoxic shelf waters in the Peruvian Oxygen Minimum Zone: Geochimica et cosmochimica acta, v. 285, pp. 207–24.

Oze, C., Bird, D. K., and Fendorf, S., 2007, Genesis of hexavalent chromium from natural sources in soil and groundwater: Proceedings of the National Academy of Sciences, v. 104, no. 16, pp. 6544–9.

Patterson, R. R., Fendorf, S., and Fendorf, M., 1997, Reduction of hexavalent chromium by amorphous iron sulfide: Environmental Science & Technology, v. 31, no. 7, pp. 2039–44.

Paulukat, C., Gilleaudeau, G. J., Chernyavskiy, P., and Frei, R., 2016, The Cr-isotope signature of surface seawater – A global perspective: Chemical Geology, v. 444, pp. 101–9.

Pereira, N. S., Vögelin, A. R., Paulukat, C. et al., 2015, Chromium isotope signatures in scleractinian corals from the Rocas Atoll, Tropical South Atlantic: Geobiology, v. 4, no. 1, p. 1–13.

Pereira, N. S., Voegelin, A. R., Paulukat, C. et al., 2016, Chromium-isotope signatures in scleractinian corals from the Rocas Atoll, Tropical South Atlantic: Geobiology, v. 14, no. 1, pp. 54–67.

Planavsky, N. J., Cole, D.B., Isson, T.T. et al., 2018, A case for low atmospheric oxygen levels during Earth's middle history: Emerging Topics in Life Sciences, p. ETLS20170161.

Planavsky, N. J., Reinhard, C. T., Wang, X. L. et al., 2014, Low Mid-Proterozoic atmospheric oxygen levels and the delayed rise of animals: Science, v. 346, no. 6209, pp. 635–8.

Reinhard, C. T., Planavsky, N. J., Robbins, L. J. et al., 2013, Proterozoic ocean redox and biogeochemical stasis: Proceedings of the National Academy of Sciences USA, v. 110, pp. 5357–62.

Reinhard, C. T., Planavsky, N. J., Wang, X. et al., 2014, The isotopic composition of authigenic chromium in anoxic marine sediments: A case study from the Cariaco Basin: Earth and Planetary Science Letters, v. 407, pp. 9–18.

Remmelzwaal, S. R. C., Sadekov, A. Y., Parkinson, I. J. et al., 2019, Post-depositional overprinting of chromium in foraminifera: Earth and Planetary Science Letters, v. 515, pp. 100–11.

Richard, F. C., and Bourg, A. C. M., 1991, Aqueous geochemistry of chromium: A review: Water Research, v. 25, no. 7, pp. 807–16.

Rickli, J., Janssen, D. J., Hassler, C., Ellwood, M. J., and Jaccard, S. L., 2019, Chromium biogeochemistry and stable isotope distribution in the Southern Ocean: Geochimica Et Cosmochimica Acta, v. 262, pp. 188–206.

Rodler, A. S., Frei, R., Gaucher, C., and Germs, G. J. B., 2016, Chromium isotope, REE and redox-sensitive trace element chemostratigraphy across the late Neoproterozoic Ghaub glaciation, Otavi Group, Namibia: Precambrian Research, v. 286, pp. 234–49.

Rudnick, R. L., and Gao, S., 2003, Composition of the continental crust: Treatise of Geochemistry, v. 3, pp. 1–64.

Saad, E. M., Wang, X. L., Planavsky, N. J., Reinhard, C. T., and Tang, Y. Z., 2017, Redox-independent chromium isotope fractionation induced by ligand-promoted dissolution: Nature Communications, v. 8, pp. 1–10.

Sander, S., and Koschinsky, A., 2000, Onboard-ship redox speciation of chromium in diffuse hydrothermal fluids from the North Fiji Basin: Marine Chemistry, v. 71, no. 1–2, pp. 83–102.

Sander, S., Koschinsky, A., and Halbach, P., 2003, Redox speciation of chromium in the oceanic water column of the Lesser Antilles and offshore Otago Peninsula, New Zealand: Marine and Freshwater Research, v. 54, no. 6, pp. 745–54.

Schauble, E., Rossman, G. R., and Taylor Jr, H. P., 2004, Theoretical estimates of equilibrium chromium isotope fractionations: Chemical Geology, v. 205, no. 1–2, pp. 99–114.

Scheiderich, K., Amini, M., Holmden, C., and Francois, R., 2015, Global variability of chromium isotopes in seawater demonstrated by Pacific, Atlantic, and Arctic Ocean samples: Earth and Planetary Science Letters, v. 423, pp. 87–97.

Schoenberg, R., Zink, S., Staubwasser, M., and von Blanckenburg, F., 2008, The stable Cr isotope inventory of solid Earth reservoirs determined by double spike MC-ICP-MS: Chemical Geology, v. 249, no. 3–4, pp. 294–306.

Scholz, F., Severmann, S., McManus, J. et al., 2014, On the isotope composition of reactive iron in marine sediments: Redox shuttle versus early diagenesis: Chemical Geology, v. 389, pp. 48–59.

Shaw, T. J., Gieskes, J. M., and Jahnke, R. A., 1990, Early diagenesis in differing depositional environments: The response of transition metals in pore water: Geochimica et Cosmochimica Acta, v. 54, pp. 1233–46.

Sial, A. N., Campos, M. S., Gaucher, C. et al., 2015, Algoma-type Neoproterozoic BIFs and related marbles in the Serido Belt (NE Brazil): REE, C, O, Cr and Sr isotope evidence: Journal of South American Earth Sciences, v. 61, pp. 33–52.

Sperling, E. A., Halverson, G. P., Knoll, A. H., Macdonald, F. A., and Johnston, D. T., 2013, A basin redox transect at the dawn of animal life: Earth and Planetary Science Letters, v. 371, pp. 143–55.

Sun, Z. Y., Wang, X. L., and Planavsky, N., 2019, Cr isotope systematics in the Connecticut River estuary: Chemical Geology, v. 506, pp. 29–39.

Toma, J., Holmden, C., Shakotko, P., Pan, Y., and Ootes, L., 2019, Cr isotopic insights into ca. 1.9 Ga oxidative weathering of the continents using the Beaverlodge Lake paleosol, Northwest Territories, Canada: Geobiology, v. 17, no. 5, pp. 467–89.

Towe, K. M., 1970, Oxygen-collagen priority and early Metazoan fossil record: Proceedings of the National Academy of Sciences of the United States of America, v. 65, no. 4, pp. 781–788.

Wang, X. L., Glass, J. B., Reinhard, C. T., and Planavsky, N. J., 2019, Species-dependent chromium isotope fractionation across the Eastern Tropical North Pacific Oxygen Minimum Zone: Geochemistry Geophysics Geosystems, v. 20, no. 5, pp. 2499–514.

Wang, X. L., Planavsky, N. J., Hull, P. M. et al., 2017, Chromium isotopic composition of core-top planktonic foraminifera: Geobiology, v. 15, no. 1, pp. 51–64.

Wang, X. L., Planavsky, N. J., Reinhard, C. T. et al., 2016a, Chromium isotope fractionation during subduction-related metamorphism, black shale weathering, and hydrothermal alteration: Chemical Geology, v. 423, pp. 19–33.

Wang, X. L., Reinhard, C. T., Planavsky, N. J. et al., 2016b, Sedimentary chromium isotopic compositions across the Cretaceous OAE2 at Demerara Rise Site 1258: Chemical Geology, v. 429, pp. 85–92.

Wei, W., Frei, R., Chen, T. Y. et al., 2018, Marine ferromanganese oxide: A potentially important sink of light chromium isotopes?: Chemical Geology, v. 495, pp. 90–103.

Wu, W. H., Wang, X. L., Reinhard, C. T., and Planavsky, N. J., 2017, Chromium isotope systematics in the Connecticut River: Chemical Geology, v. 456, pp. 98–111.

Zink, S., Schoenberg, R., and Staubwasser, M., 2010, Isotopic fractionation and reaction kinetics between Cr(III) and Cr(VI) in aqueous media: Geochimica Et Cosmochimica Acta, v. 74, no. 20, pp. 5729–45.

# Elements in Geochemical Tracers in Earth System Science

## Timothy Lyons

*University of California*

Timothy Lyons is a Distinguished Professor of Biogeochemistry in the Department of Earth Sciences at the University of California, Riverside. He is an expert in the use of geochemical tracers for applications in astrobiology, geobiology and Earth history. Professor Lyons leads the 'Alternative Earths' team of the NASA Astrobiology Institute and the Alternative Earths Astrobiology Center at UC Riverside.

## Alexandra Turchyn

*University of Cambridge*

Alexandra Turchyn is a University Reader in Biogeochemistry in the Department of Earth Sciences at the University of Cambridge. Her primary research interests are in isotope geochemistry and the application of geochemistry to interrogate modern and past environments.

## Chris Reinhard

*Georgia Institute of Technology*

Chris Reinhard is an Assistant Professor in the Department of Earth and Atmospheric Sciences at the Georgia Institute of Technology. His research focuses on biogeochemistry and paleoclimatology, and he is an Institutional PI on the 'Alternative Earths' team of the NASA Astrobiology Institute.

## About the Series

This innovative series provides authoritative, concise overviews of the many novel isotope and elemental systems that can be used as "proxies" or "geochemical tracers" to reconstruct past environments over thousands to millions to billions of years – from the evolving chemistry of the atmosphere and oceans to their cause-and-effect relationships with life.

Covering a wide variety of geochemical tracers, the series reviews each method in terms of the geochemical underpinnings, the promises and pitfalls, and the "state-of-the-art" and future prospects, providing a dynamic reference resource for graduate students, researchers and scientists in geochemistry, astrobiology, paleontology, paleoceanography, and paleoclimatology.

The short, timely, broadly accessible papers provide much-needed primers for a wide audience – highlighting the cutting-edge of both new and established proxies as applied to diverse questions about Earth system evolution over wide-ranging time scales.

Cambridge Elements ☰

# Elements in Geochemical Tracers in Earth System Science